Little Astronaut
and
His Aliens Buddy

Coloring Book
for Kids

ColorFun
PUBLISHING

Please help to leave your review about this book.
With your feedback, we'll make
our next books even better!
Thanks, and enjoy your coloring journey.

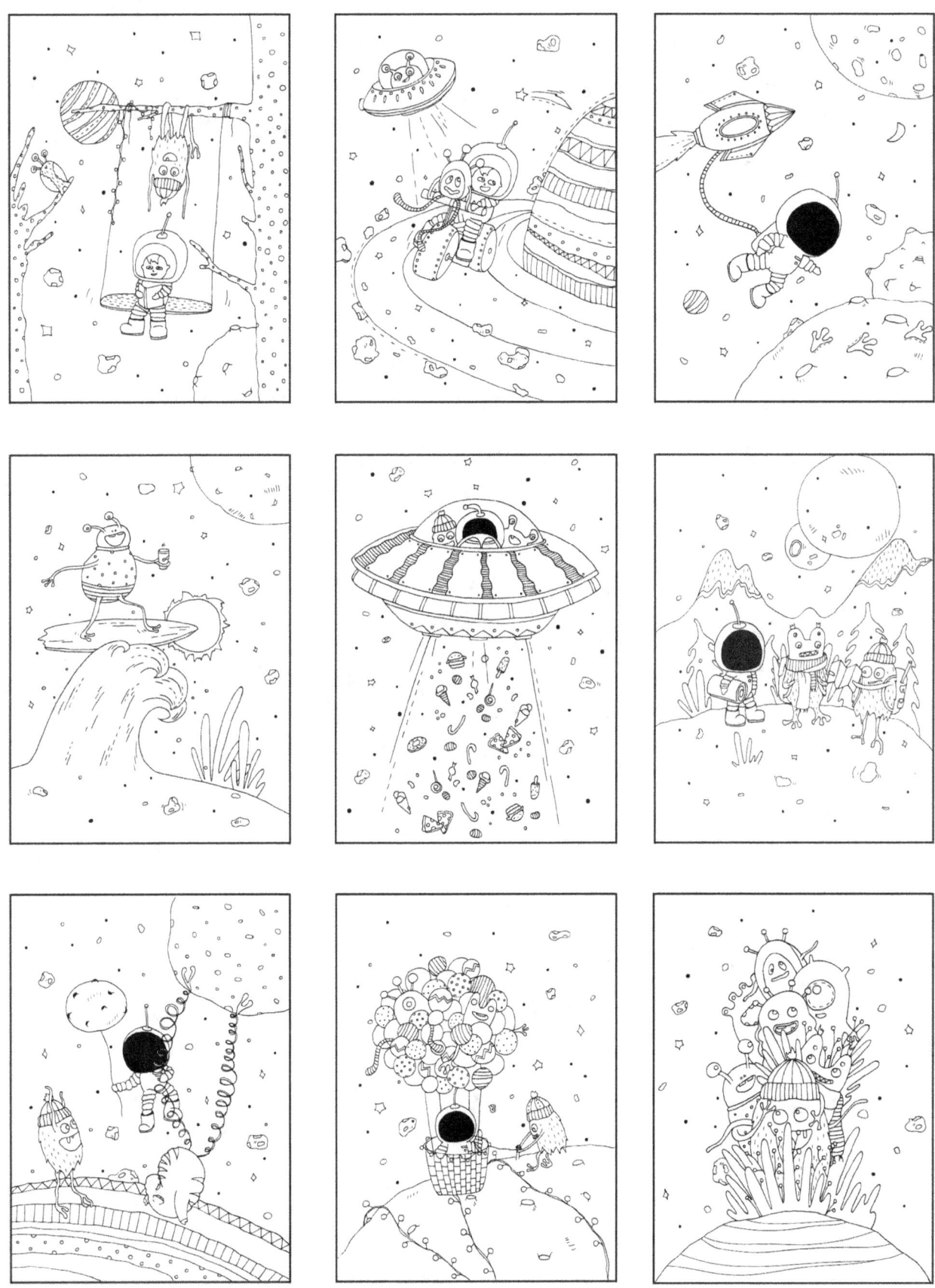

Color test page

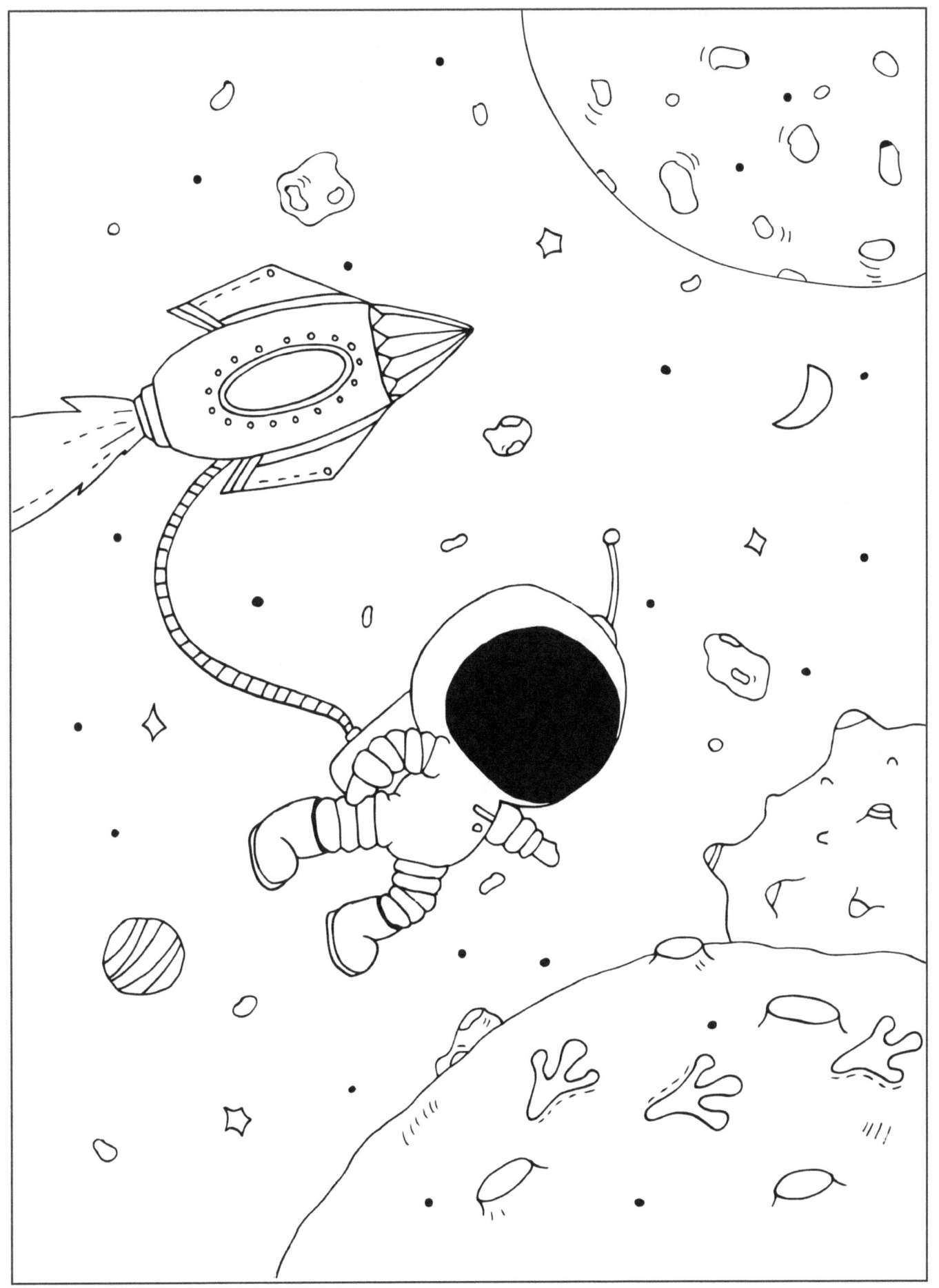

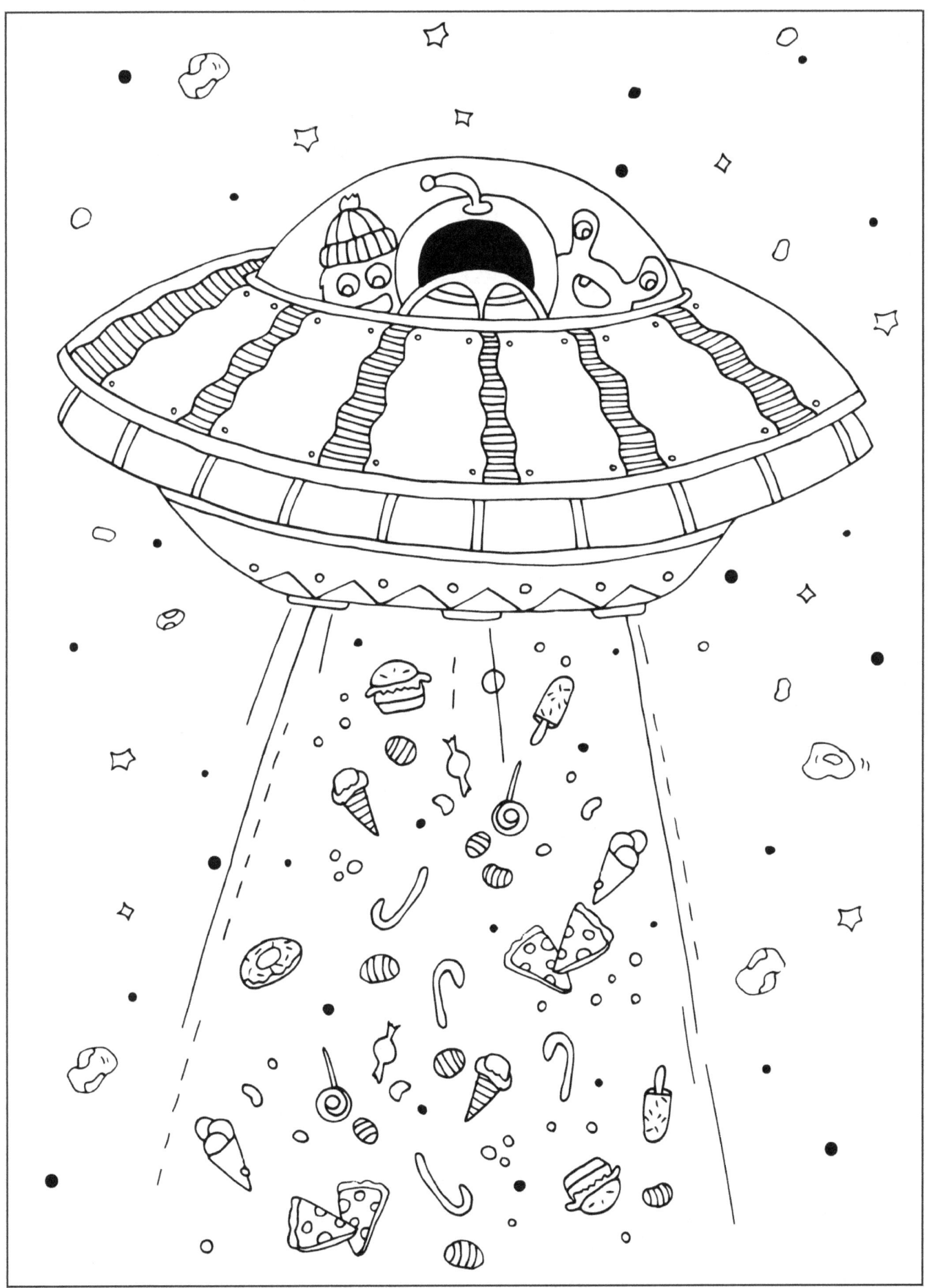

www.ingramcontent.com/pod-product-compliance
Lightning Source LLC
Chambersburg PA
CBHW080855220526
45467CB00008B/2514